Seed Dispersal

by William J. Beal

SEED DISPERSAL.

CHAPTER I.

HOW ANIMALS GET ABOUT.

1. Most of the larger animals move about freely.--When danger threatens, the rabbit bounds away in long jumps, seeking protection in a hollow tree, a log, or a hole in the ground. When food becomes scarce, squirrels quickly shift to new regions. Coons, bears, skunks, and porcupines move from one neighborhood to another. When the thickets disappear and hunters abound, wild turkeys and partridges retreat on foot or by wing. When the leaves fall and the cold winds blow, wild geese leave the lakes in secluded northern homes, and with their families, reared during the summer, go south to spend the winter. Turtles swim from pond to pond or crawl from the water to the sand bank, where they lay and cover their eggs. Fishes swim up or down the creek with changing seasons, or seek deep or shallow water as their needs require. Beetles and butterflies, when young, crawl about for food and shelter, and when older use their wings in going long distances.

These examples only serve to recall to mind what every boy or girl knows and has known ever since he can remember--that most animals move about whenever they want to, or whenever other animals will let them.

2. Some animals catch rides in one way or another.--Some small animals, like lice, ticks, and tiny spiders, walk slowly and only for short distances. If, because of scarcity of food, they are suddenly seized with the desire to move for a long distance, what are they to do? On such occasions ticks and lice watch quietly the first opportunity, catch on to the feet of birds or flying insects or other animals which may happen to come their way, and, like a boy catching on to a farmer's sleigh, ride till they get far enough, then jump off or let go, to explore the surrounding country and see whether it is fit to live in. If for some reason a spider grows dissatisfied and wants to leave the home spot, she climbs to the top of some object and spins out a fine, long web; this floats

in the air, and after a while becomes so long and light that the wind will bear the thread and the spinner for a considerable distance, no one knows how far. These facts about lice and spiders show how wingless insects can go long distances without wings of their own.

How is it with plants? The woods, fields, marshes, roadsides ever abound with interesting objects provided with strange devices waiting to be studied by inquisitive girls and boys in and out of school, and this finding out of nature's puzzles is one of the deepest pleasures of life.

How quickly a mould attacks and creeps or spreads through a basin of berries every one knows. The mould is as much a plant as the bush that produced the berries; it comes from a small spore, which takes the place of a bud or sprout or seed. The decay of a tree begins where a limb or root has been injured, and whether the timber is living or dead, this decay results from the growth of some one or more low forms of plant life which enter the timber in certain places and slowly or quickly penetrate and affect other portions more or less remote.

CHAPTER II.

PLANTS SPREAD BY MEANS OF ROOTS.

3. Fairy rings.--Several low forms of plant life, such as Marasmius oreades, Spathularia flavida, and some of the puffballs, start in isolated spots in the grass of a lawn or pasture, and spread each year from a few inches to a foot or more in every direction, usually in the form of a circle; at the end of fifteen years some of these circles acquire a diameter of fifteen to twenty feet or more. These are known as fairy rings. Before science dispelled the illusion they were believed to have been the work of witches, elves, or evil spirits, from which arose the name.

Several kinds of lichens and mosses and the like, growing on the barks of trees, fence boards, and low ground, spread slowly in the manner of fairy rings.

However, the spreading is not always a slow, creeping process, for sometimes these low plants spread over an incredible distance in a short space of time. In some instances they appear suddenly almost anywhere, and at any season of the year. They are all minute and exist in countless numbers, and their devices for securing wide dispersion are so various as to entitle them to first rank in this respect. Some send off spores with a sharp puff, as if shot from a little gun. Some of these spores float on water, and some are sticky and thus gain free rides. It is not at all improbable that some are carried by the winds across oceans and continents.

It is well known that many of the lower species of plants are more widely distributed over the earth than most of the higher plants. Every cloud from a ripe puffball consists of thousands of spores started on the wings of the wind for an unknown journey. Their habits are not past finding out, but to examine them a person needs a good microscope. Most of them have no special common name, and with one or two exceptions further mention of the mode of distribution of this fascinating portion of plant life cannot here be made.

In our botanic garden was planted a patch six feet across of what is known as Oswego tea, bee balm, or red-flowered bergamot, an interesting plant with considerable beauty. It grew well for a year, the next year it failed to some extent, and on the third most of the plants died, or nearly died, excepting the spreading portion all around the margin. This is a fairy ring of another type, and represents a very slow mode of travel. As further illustrations of this topic study common yarrow, betony, several mints, common iris, loosestrife, coreopsis, gill-over-the-ground, several wild sunflowers, horehound, and many other perennials that have grown for a long time without transplanting.

The roots of plants are seldom much observed, because they are out of sight. In soft ground the roots of the common or black locust extend from twenty to forty feet in each direction, and almost anywhere along these roots buds may appear, and a shoot spring up and become a tree.

This peculiarity is worth as much to locusts in the matter of spreading as though the parent trees were able to move about. A number of kinds of poplars and willows, ailanthus, some of the elms, ashes, sweet potatoes, milkweeds, Canada thistles, and others behave in a similar manner. Little bits of Canada-thistle root half an inch long may send forth buds, and each bud grow to be an independent plant.

Roots have a peculiarity not usually known. They stretch out and crook about here and there, penetrating the crevices of the soil wherever there is the least chance, and the matured portions begin to shorten, reminding one somewhat of an angleworm when one end has been stepped on. By this shortening process the top or crown of a dandelion or plantain is pulled down beneath the surface of the ground.

4. How nature plants lilies.--Lilies grow from bulbs which are planted six inches beneath the surface. Do you know how nature plants them? A seed starts and becomes a small plant on the surface of the leaf mould or a little beneath; little roots push downward and to right and left; and later, after getting a good hold below with numerous branchlets, the slender roots shorten and tug away at the tiny bulb above, as much as to say, "Come down a little into mother earth, for cold winter is approaching and there will be danger from frost." The young bulb is drawn down an inch more or less, the slender roots perish with the growing year, but the bulb is preserved. The seedling was well planned; for while it had yet tender leaves during its first year, starch and protoplasm were stored up in the thickened scales of the bulb. During the second spring some of this food in store is used to send down another set of slender roots with the message to gather in more water, potash, phosphorus, nitrogen, and other substances to help grow a larger bulb. In late summer and autumn the new roots contract and pull away at the greater bulb, and down it goes into the ground another inch or so. I have a theory as to how it finally comes to be drawn down just deep enough and no more, but I will not venture to give it. This process is repeated from year to

year till the proper depth is reached for preserving the full-grown bulb. And this is the way nature plants bulbs.

In a similar manner young slender roots well anchored in the soil, at or near the close of the growing season, pull downward and outward large numbers of bulblets that form around a parent bulb of some kinds of leeks, tulips, star-of-bethlehem, globe hyacinth, and monkshood. The pull of the roots is much greater to one side than downward, because most of the longest roots extend sidewise. Marilaun reports that a certain lawn in Vienna was mown so frequently that tulips could not go to seed, but after twenty years, from a very few bulbs planted near each other, a space twenty paces in diameter was well covered by tulips. And this is one way tulips travel, slow and sure.

5. Roots hold plants erect like ropes to a mast.--Did you ever lift vines of cucumbers, squashes, and the like, where they had rooted at the joints, and observe how forlorn they looked after the operation, with leaves tipped over, unable to remain erect? While growing, the stem zigzags or winds about more or less, and thus enables it to hold the leaves erect; besides, the tendrils catch on to weeds and curl up tight, and the roots at the joints are drawn taut on each side after the manner mentioned above, and act like ropes to a mast to hold the stem in its place, and thus help to hold the leaf above erect.

6. How oaks creep about and multiply.--Oaks come from acorns; everybody knows that. The nuts are produced in abundance, and those of the white oak send out pretty good tap roots on the same year they fall. Some of the nuts roll down the knoll or are carried about by squirrels or birds, as mentioned elsewhere. Let me tell you one thing that I discovered the white oaks were doing in the sand of the Jack-pine plains of Michigan. In dry weather the dead grass, sticks, and logs are often burned, which kills much or all that is growing above ground. In this way little maples, ashes, witch-hazels, willows, huckleberries, blackberries, sweet ferns, service berries, aspens, oaks, and others are often killed back, but afterward sprout up again and again, and, after repeated burnings, form each a large rough mass popularly known as a grub. The grubs of the oak are well known; the large ones weighing from 75

to 100 pounds each. To plow land where grubs abound requires a stout plow and several pairs of horses or oxen.

A small white oak, after it has been many times killed to the ground, dies in the middle and sprouts at the margins, and finally the main root perishes, and two roots, with branches a little distance apart, support each a cluster of stems above ground.

There can be no doubt that young oak trees slowly move in this manner from one place to another. If in fifty years we have two distinct grubs or branches, three or four feet apart, where the connecting part has finally died out, I see no reason why in another fifty years each one of the two may not again have spread and divided, giving us at least four grubs, or clusters of sprouts, all originally coming from one acorn; and so the matter might go on. This is slow traveling, I admit, but there is nothing to hinder nature from taking all the time she wants.

CHAPTER III.

PLANTS MULTIPLY BY MEANS OF STEMS.

7. Two grasses in fierce contention.--In growing a lawn at the Michigan Agricultural College, a little Bermuda grass was scattered with June grass, and the struggle has been most interesting. In the spring and for six weeks in autumn, when moisture usually abounds and the weather is cool, June grass thrives and little else is seen. In the dry, hot weeks of July and August, June grass rests and the Bermuda, which continues to spread, assumes control of the lawn, with but little of the June grass in sight. Each struggles for possession and does the best it can, and to some extent one supplements the other, with the result that at all times from spring to fall there is a close mat of living green which delights the eye and is pleasant to the feet that tread upon it. In soft ground, with plenty of room, a bit of quick or quack grass, or Bermuda, will extend in a year three to five feet or more in one direction.

June grass, quick grass, Bermuda grass, redtop, and white clover, wherever opportunity offers, spread by means of jointed stems, creeping and rooting at every joint on the surface of the ground or a little way below. These are not roots at all, but true stems somewhat in disguise. Here may also be mentioned, as having similar habit, artichokes, peppermint, spearmint, barberry, Indian hemp, bindweed, toadflax, matrimony vine, bugle-weed, ostrich fern, eagle fern, sensitive fern, coltsfoot, St. John'swort, sorrel, great willow-herb, and many more.

8. Runners establish new colonies.--The spreading of strawberries by runners must be familiar to every observer. In 1894 a student reported that a wild strawberry plant in the botanic garden had produced in that year 1230 plants. Weeds were all kept away, the season was favorable, the soil sandy; but on one side, within a foot and a half, progress was checked by the presence of a large plant of another kind. The multiplication of this plant by seeds, in addition to that by runners, would have covered a still greater area of land. Other plants with runners much like the strawberry are: several kinds of crowfoot, barren strawberry, cinquefoil, strawberry geranium, and orange hawkweed. Plants of the star cucumber, one-seeded cucumber, grapes, morning-glories, and others, spread more or less over bushes or over the ground, and are thus enabled to scatter seeds in every direction.

9. Branches lean over and root in the soil.--A black raspberry grows fast in the ground and has to stay in one spot for life. It has neither legs, feet, nor wings, and yet it can travel. The bush takes deep root and spreads out its branches, which are sometimes ten feet or more in length; the tips of these branches curve over to the ground six feet away, and finally take root; from these roots new colonies are formed, five to twenty in a year from one bush.

True, the old roots do not get far, and the new plants only get about six feet in one season, but they have made some progress. This is rather slow locomotion, you say; but let us look a little farther, remembering that a seed is a little plant packed ready for transportation. This second mode of spreading will be described on a future page.

10. Living branches snap off and are carried by water or wind.--Some trees and shrubs among the willows are called snap-willows, because their branches are very brittle; on the least strain from wind, rain, sleet, or snow, the smaller branches snap off near the larger branches or the main trunk, and fall to the ground. At first thought this brittleness of the wood might seem to be a serious defect in the structure of the tree or shrub, although they seem to produce branches enough for their own use.

But the branches which are strewn all around after a storm often take root in the low ground where they fall; some of them are carried down stream by the current, and, lodging on the shore below, produce new trees or bushes. During the winter of 1895 and 1896 a group of seven white willows, near a brook on the campus of the Michigan Agricultural College, was at one time loaded with sleet. There was considerable snow on the ground, which, of course, was covered with an icy crust. In a little while the sleet melted from the fallen branches strewn about, and a moderate breeze then drifted the smallest of the twigs in considerable numbers over the icy snow. Some of these were found thirty rods distant from the parent trees--not down stream in the valley of the brook, but up the stream. Had not the low ground been covered with a dense growth of grass, some of these branches might have started new trees where the wind had left them.[1]

[Footnote 1: C. D. Lippincott believes that this is a provision of nature to dispose of the now unnecessary branchlets without leaving a knot. Plant World, Vol. I, p. 96.]

The branches on slow-growing limbs of cottonwood and large-toothed aspen are much enlarged at the nodes, and at these places are brittle, often separating from the tree and breaking up into pieces. Under a small cottonwood were picked up a bushel or more of such limbs, all yet alive. These trees are common on low land, and, like snap-willows, the severed twigs may find a chance to grow on moist soil.[2]

[Footnote 2: The brittle branches of salix were noticed by the author in Bull. Torr. Bot. Club, Vol. IX (1883), p. 89.]

In a greenhouse a potted plant of Selaginella emiliana(?) was placed on the bench near the aisle, where it was often brushed by people in passing. Small branches, not being firmly attached, were frequently broken from the main plant and fell upon the moist sand, where they rooted in abundance.

CHAPTER IV.

WATER TRANSPORTATION OF PLANTS.

11. Some green buds and leaves float on water.--Loosely floating on slow streams of the northern states, in water not the purest, may often be found the common bladderwort, Utricularia vulgaris, producing in summer a few yellow flowers on each stem, rising from six to twelve inches above the water. The lax, leafy branches in the water are from six inches to a foot long. The leaves, or thread-like branches, are about half an inch long, more or less, and several times divided.

Scattered about are large numbers of flattened scales, or bladders, sometimes one-sixth of an inch long, which give the plant one of its names. For a long time the bladders were thought to serve merely as life-preservers; it was supposed that they were constructed to keep the plant from sinking to the bottom. In reality these bladders help preserve the plant in another sense, by catching and killing large numbers of minute animals, on which the plant lives in part. The tips of the stems at all times of the year are rather compact, made up of young leaves and stems, and in the middle of the summer, as well as at other times, many may be seen severed from the parent plant, floating in the water, ready to accept the assistance of any favorable current or breeze and start out for homes of their own to found new colonies. These olive-green tips, or buds, vary much in size, but the largest are the size of the end of one's little finger. Late in autumn or early winter, when cold threatens, the tender buds contract a little, and, having thus become heavier than water, slowly go to the bottom to spend the winter safely protected in the soft mud. All the plant perishes except these buds. With the lengthening days of spring the melting ice disappears, and genial sunshine gives notice to the dormant buds that it is safe to come out again. The buds begin to expand, become lighter than water, and are soon seen spreading out at the surface and producing branches and leaves. Ducks and other water-fowl not infrequently carry some of these wet buds sticking to their feathers or legs.

In this connection the following plants may be examined from time to time: Lemna, Wolffia, Anacharis (Elodea), Myriophyllum, Cabomba, and several species of Potamogeton. I have seen the leaves of lake cress, Nasturtium lacustre, often spontaneously separate from the stem, possibly carrying at the base the rudiments of a small bud, which draws on the floating leaf for nourishment and produces a small plant near its base. These plants, floated and nourished by the mother leaf, may drift down a creek or across a pond and establish new settlements. In a similar manner behave leaves of the following, and perhaps others: Cardamine pratensis, horse-radish, celandine, some water lilies, and other plants not grown in wet land.

Gardeners often propagate certain species by placing leaves on wet sand or mud, when buds spring from the margins of the leaves or from some other portion.

One of the buttercups, Ranunculus multifidus, and very likely others, spread over the mud by producing runners, much after the manner of a strawberry plant. If, as in case of a freshet, the plants should be covered with water, they show their enterprise by taking advantage of the "tide"; some of the runners are quickly severed, and are then at liberty to go as they please.

12. Fleshy buds drop off and sprout in the mud.--One of the loosestrifes, Lysimachia stricta, a plant growing in bogs, besides reproducing itself by rootstocks and seeds, bears fleshy buds half an inch long, which separate from the stems and take root in the mud near the parent plant, or often float to another spot. The buds on the stems of Cicuta bulbifera develop into small

bulbs, which readily separate from the plant. They then float on the water and produce new plants. The tiger lily also produces bulblets, which scatter about and promptly take root. Every person of good understanding must have heard or read about seeds carried by ocean currents or transported by lake, pond, creek, or by muddy current, during, and after, a shower of rain; in most of these the wind is also a prominent factor. Many seeds and fruits, in some cases parts, and even the whole, of plants seem to be purposely designed for this mode of travel, while an innumerable host of others occasionally make use of it, although it may seem from their structure and place of growth that they were made especially to be transported by the wind or by some animal. As has been seen in examples previously mentioned, one portion of a plant is transported in one way, and another portion by one or two other methods.

13. Seeds and fruits as boats and rafts.--An excellent place in which to begin investigating this part of the subject is to pay a visit to the flats of a creek or river late in autumn or in the spring, after the water has retired to its narrow channel, and examine piece after piece of the rubbish that has been lodged here and there against a knoll or some willows, a patch of rushes or dead grass. We are studying the different modes by which plants travel. In the driftwood may be found dry fruits of the bladder nut, brown and light, an inch and a half in diameter. See how tough they are; they seem to be perfectly tight, and even if one happens to have a hole punched in its side, there are probably two cells that are still tight, for there are three in all. Within are a few seeds, hard and smooth. Why are they so hard? Will it not be difficult for such seeds to get moist enough and soft enough to enable them to germinate? The hard coats enable the seeds to remain uninjured for a long time in the water, in case one or two cells of the papery pods are broken open; and after the tough pod has decayed and the seeds have sunken to the moist earth among the sticks and dead leaves, they can have all the time they need for the slow decay of their armor. Sooner or later a tiny plant is likely to appear and produce a beautiful bush. Engineers are boasting of their steel ships as safe and not likely to sink, because there are several compartments each in itself water-tight. In case of accident to one or two

chambers, the one or two remaining tight will still float the whole and save the passengers.

I wonder if the engineers have not been studying the fruit of the bladder nut? But this is not all. Many of the dry nuts hang on all winter, or for a part of it, rattling in the wind, as though loath to leave. Some of them are torn loose, and in winter there will be a better chance than at any other time for the wind to do the seeds a favor, especially when there is snow on the ground, for then they will bound along before the breeze till something interrupts them.

Here among the rubbish are some shriveled wild grapes also. As we shall see elsewhere, their best scheme is to be eaten by certain birds, which do not digest their bony seeds; but in case some of them are left there is another mode of travel, not by wings of a bird, but by floating on water.

Clean grape seeds sink at once, but covered by the dry skin and pulp, they float. In a similar manner the dry seeds of several dogwoods are eaten for the pulp by birds, but in case any are left they behave after the manner of grapes.

14. Bits of cork around the seeds prevent them from sinking.--Narrow-leaved dock is a prominent weed, and is especially at home on river bottoms and on low land that is flooded once in a while.

Did you ever wonder what could be the object of a round, spongy tubercle on the outside of each of these sepals which hold the ripened seed closely? I did not know their use for a long time, but now think I have discovered their meaning. They are not exactly life-preservers, but the next thing to it. The naked, seed-like fruit, the shape of the fruit of buckwheat, sinks at once when free from everything else, but with the dry calyx still attached, it floats with the stream.

15. An air-tight sack buoys up seeds.--Here are several dry fruits of sedges--plants looking considerably like grasses. There are a good many kinds, and

most of them grow in wet places. The seed-like fruit of those we examine are surrounded each by a sack which is considerably too large for it, as one would be likely to say, but in reality it serves to buoy the denser portion within, much after the plan of the bladder nut. In some instances the sack is rather small, but a corky growth below the grain helps to buoy it on water.

Sedges that grow on dry land usually have the sack fitted closely, instead of inflated, and the whole mass sinks readily in water. Now we see the probable reason why the sack is inflated in some species of sedges and not in others.

Here are some small, seed-like fruits, achenes, not likely to be recognized by every one. They belong to the arrowhead, Sagittaria, found in shallow ponds or slow streams. They are flattened, and on one edge, or both, and at the apex is a spongy ridge. Very likely, by this time, the reader has surmised that this serves the purpose of a raft to float the small seed within, which would sink at once if separated from the boat that grew on its margins. In this connection may be studied achenes of water plantain, Alisma, bur reed, cat-tail flag, arrow grass, burgrass, numerous pondweeds, several buttercups, the hop, nettles, wood nettle, false nettle, cinquefoil, avens, ninebark, buttonbush, and in fact a large number and variety of plants usually found on river bottoms.

One of the lyme grasses, Elymus Virqinicus, is a stiff, short grass, growing along streams. Each spikelet with its chaff adheres to two empty glumes, stout, thick, and spongy, which make a safe double boat for transportation down stream whenever the water is high enough. The grains of rice-cut grass, grown in ditches and spring brooks, sink if separated, but in the chaff, as they fall when ripe, they are good floaters.

In the driftwood, which we still have under consideration, are some fruits of maple, beech, oak, tulip tree, locust, and basswood. Maples are well scattered by the wind, but these seed-like fruits have taken to the water, and a few still retain vitality. An acorn, while yet alive, sinks readily, and is not suited for water navigation, unless by accident it rides on some driftwood.

The fruits of the tulip tree, locust, and basswood behave well on the water, as though designed for the purpose, though we naturally, and with good reason, class them with plants usually distributed by wind.

16. Fruit of basswood as a sailboat, and a few others as adapted to the water.--In spring, when the bracts and fruits of the basswood are dry and still hanging on the tree, if a quantity of them are shaken off into the water which overflows the banks of a stream, many of these, as they reach the water, will assume a position as follows: The nuts spread right and left and float; the free portion of the bract extends into the water, while the portion adhering to the peduncle rises obliquely out of the water and serves as a sail to draw along the trailing fruit. After sailing for perhaps fifteen minutes, the whole bract and stem go under water, the nuts floating the whole as they continue to drift with the wind.

Noticeable among seeds in the flood wood are some of the milkweeds, which every one would say at a glance were especially fitted for sailing through the air, aided by their numerous long, silky hairs. These hairs are no hindrance to moving by water. I discovered one little thing in reference to the seed which makes me think the Designer intended it should to some extent be carried by water. The flat seed has a margin, or hem, which must be an aid to the wind in driving it about; but this margin is thickened somewhat by a spongy material.

With the margin it floats, without it the seed sinks in fresh water. A few cranberries were found in the driftwood. These contain considerable air in the middle, near where the seeds are placed, as though the air was intended to support them on top of water.

These berries are colored and edible--qualities that attract the birds. And here we find in several places the bulblets of a wild garlic, Allium Canadense, which grows on the river bottom. These bulblets are produced on top of the stem with the flowers, and float on the water. The seeds of the white water lilies, and yellow ones also, by special arrangement float about on the water with the current or the wind. The coffee tree grows rather sparingly along some of the streams, and on moist land as far north as Clinton County, Michigan. The stout, hard pods are three to four inches long, one and one-quarter to one and one-half inches wide, and one-half inch thick. The very hard seeds are surrounded with sweet pulp, which most likely made it an inducement for some of our native animals to devour them and thus transport the undigested seeds to remote localities. The pods often remain on the trees all winter, and when dry, will float on the water of overflowed streams without any injury resulting to the hard seeds. By themselves the seeds sink at once.

CHAPTER V.

SEEDS TRANSPORTED BY WIND.

17. How pigweeds get about.--In winter we often see dead tops of lamb's-quarters and amaranths--the smooth and the prickly pigweeds--still standing where they grew in the summer. These are favorite feeding grounds for several kinds of small birds, especially when snow covers the ground.

Many of the seeds, while still enclosed in the thin, dry calyx, and these clustered on short branches, drop to the snow and are carried off by the wind. Notwithstanding the provision made for spreading the seeds by the aid of birds and the wind, the calyx around each shiny seed enables it to float also; when freed from the calyx, it drops at once to the bottom. Many kinds of dry fruits and seeds in one way or another find their way during winter to the surface of the ice-covered rivers. When the rivers break up, the seeds are carried down stream, and perhaps left to grow on dry land after the water has retired. Most of the commonest plants, the seeds of which are usually transported by water, are insignificant in appearance and without common names, or with names that are not well understood. This is one reason for omitting the description of others which are ingeniously fitted in a great variety of different ways for traveling by water.

18. Tumbleweeds.--Incidentally, the foregoing pages contain some account of seeds and fruits that are carried by the aid of wind, in connection with their distribution by other methods; but there are good reasons for giving other examples of seeds carried by the wind. There is a very common weed found on waste ground and also in fields and gardens, which on good soil, with plenty of room and light, grows much in the shape of a globe with a diameter of two to three feet. It is called Amaranthus albus in the books, and is one of the most prominent of our tumbleweeds. It does not start in the spring from seed till the weather becomes pretty warm. The leaves are small and slender, the flowers very small, with no display, and surrounded by little rigid, sharp-pointed bracts. When ripe in autumn, the dry, incurved branches are quite stiff; the main stem near the ground easily snaps off and leaves the light ball at the mercy of the winds. Such a plant is especially at home on prairies or cleared fields, where there are few large obstructions and where the wind has free access.

The mother plant, now dead, toiled busily during the heat of summer and produced thousands of little seeds. The best portion of her substance went to produce these seeds, giving each a portion of rich food for a start in life and wrapping each in a glossy black coat. Now she is ready to sacrifice the rest of her body to be tumbled about, broken in pieces, and scattered in every direction for the good of her precious progeny, most of whom will find new places, where they will stand a chance the next summer to grow into plants. Sometimes the winds are not severe enough or long enough continued, and these old skeletons are rolled into ditches, piled so high in great rows or masses against fences that some are rolled over the rest and pass on beyond. Occasionally some lodge in the tops of low trees, and many are entangled by straggling bushes. In a day or two, or in a week, or a month, the shifting wind may once more start these wrecks in other directions, to be broken up and scatter seeds along their pathway.

During the Middle Ages in southern Egypt and Arabia, and eastward, a small plant, with most of the peculiarities of our tumbleweed just described, was often seen, and was thought to be a great wonder. It was called the "rose of

Jericho," though it is not a rose at all, but a first cousin to the mustard, and only a small affair at that, scarcely as large as a cabbage head. A number of other plants of this habit are well known on dry plains in various parts of the world; one of the most prominent in the northern United States is called the Russian thistle, which was introduced from Russia with flaxseed. In Dakota, often two, three, or more grow into a community, making when dry and mature a stiff ball two to three feet or more in diameter.

One of our peppergrasses, Lepidium intermedium, sometimes attains the size and shape of a bushel basket; when ripe, it is blown about, sowing seeds wherever it goes. The plants of the evening primrose sometimes do likewise, also a spurge, Euphorbia [Preslii] nutans, a weed a foot to a foot and a half high.

Low hop clover, an annual with yellow flowers, which has been naturalized from Europe, has developed recently on strong clay land into a tumbleweed six inches in diameter. The tops of old witch grass, Panicum capillare, and hair grass, Agrostis hyemalis, become very brittle when ripe, and snap from the parent stem and tumble about singly or in masses, scattering seeds by the millions. I have seen piles of these thin tops larger than a load of hay where they had blown against a grove of trees, and in some cases many were caught in the tops of low trees.

Bug seed and buffalo bur are tumbleweeds. In autumn the careful observer with an eye to this subject will be rewarded by finding many other plants that behave more or less as tumbleweeds. Especially is this the case on prairies. These are annuals, and perish at the close of the growing season. There are numerous other devices by which seeds and fruit secure transportation by the wind.

19. Thin, dry pods, twisted and bent, drift on the snow.--The common locust tree, Robinia Pseudacacia, blossoms and produces large numbers of thin, flat pods, which remain of a dull color even when the seeds are ripe. The pods of the locust may wait and wait, holding fast for a long time, but nothing comes

to eat them. They become dry and slowly split apart, each half of the pod usually carrying every other seed. Some of the pods with the seeds still attached are torn off by the wind and fall to the ground sooner or later, according to the force of the wind. Each half-pod as it comes off is slightly bent and twisted, and might be considered a "want-advertisement" given to the wind: "Here I am, thin, dry, light and elastic, twisted and bent already; give me a lift to bear these precious seeds up the hill, into the valley, or over the plain."

And the wind is sure to come along, a slight breeze to-day tossing the half-pod a few feet, leaving it perhaps to be again and again moved farther forward. The writer has seen these half-pods transported by this means more than a block. But many of the pods stick to the limbs till winter comes. Then a breeze tears off a few pods and they fall on the snow, which has filled up all the crevices in the grass and between the dead leaves and rubbish. Each half-pod, freighted with every other seed, is admirably constructed; like an ice boat, it has a sail always spread to the breeze. In this way there is often nothing to hinder some of the seeds from going a mile or two in a few minutes, now and then striking some object which jars off a seed or two. The seeds are very hard, and no doubt purposely so, that they may not be eaten by insects or birds; but once in moist soil, the covering slowly swells and decays, allowing the young plant to escape. Thus the locust seeds are provided with neither legs, wings, fins, nor do they advertise by brilliant hue and sweet pulp; but they travel in a way of their own, which is literally on the wings of the wind.

20. Seeds found in melting snowdrifts.--It will interest the student of nature to collect a variety of seeds and dry fruits, such as can be found still on the trees and other plants in winter, and try some of them when there is snow on the ground and the wind blows, to see how they behave. Again, when the first snow banks of the early winter are nearly gone, let him collect and melt a quantity of snow and search for seeds. By this means he can see, as he never saw before, how one neighbor suffers from the carelessness of another.

21. Nuts of the basswood carried on the snow.--Here are some notes concerning the distribution of the spherical nuts of basswood. The small clusters of fruit project from a queer bract which remains attached before and after falling from the tree.

This bract, when dead, is bent near the middle and more or less twisted, with the edges curving toward the cluster of nuts. From two to five nuts about the size of peas usually remain attached till winter, or even a few till spring. This bract has attracted a good deal of attention, and for a long time everybody wondered what could be its use. We shall see. The cluster of nuts and the bract hang down, dangling about with the least breath of wind, and rattling on the trees because the enlarged base of the stem has all broken loose excepting two slender, woody threads, which still hold fast. These threads are of different degrees of strength; some break loose after a few hard gales, while others are strong enough to endure many gales, and thus they break off a few at a time. The distance to which the fruit can be carried depends on the form of the bract, the velocity of the wind, and the smoothness of the surface on which the fruit falls. When torn from the tree the twist in the bract enables the wind to keep the cluster rapidly whirling around, and by whirling it is enabled to remain longer suspended in the air and thus increase the chances for a long journey. In throwing some of these from a third-story window, it was found that a bract with no fruit attached would reach the ground sooner than a bract that bore from two to four solid nuts. The empty or unloaded bracts tumble and slide through the air endwise, with nothing to balance them or steady their descent, while the fruit on other bracts holds them with one side to the air, which prolongs their descent. The less a loaded bract whirls, the faster its descent, and the more a bract whirls when the wind blows, the farther it is carried. The bract that is weighted with a load of fruit acts as a kite held back by a string, and when in this position the wind lifts the whole as well as carries it along. Before snow had fallen in 1896, by repeated moves on a well-mowed lawn, fruit and bracts were carried about two hundred feet, while with snow on the ground the distance was almost unlimited, excepting where there were obstructions, such as bushes and fences. When there is a crust on the snow and a good wind, the conditions are almost perfect. Over the snow the wind drives the bracts, which drag along the branch of fruit much as a sail propels a boat. The curving of the edges of the bract toward the fruit enables the wind to catch it all the better, and to lift it more or less from the snow. With changes in the direction of the wind, there is an opportunity for the fruit of a single tree, if not too much crowded by others, to spread in all directions. After watching these maneuvers, no one could doubt the object of the bent bracts of the basswood, and as these vary much in length and width and shape on different trees, it would seem that perhaps nature is still experimenting with a view to finding the most perfect structure for the purpose.

About one hundred and thirty paces west of the house in which I live stand two birch trees. One windy winter day I made some fresh tracks in the snow near my house, and within a few minutes the cavities looked as though some one had sprinkled wheat bran in them, on account of the many birch seeds there accumulated.

Other fruits in winter can be experimented with, such as that of box elder, black ash, birches, tulip tree, buttonwood, ironwood, blue beech, and occasionally a maple.

22. Buttonwood balls.--Nature seems to have no end of devices for sowing seeds to advantage. Here is one which always interests me. The fruit of the buttonwood, or sycamore, which grows along streams, is in the form of balls an inch and a half in diameter. These balls grow on the tops of the highest branches, and hold on into winter or longer. The stems are about two inches long, and soon after drying, through the action of the winds, they become very flexible, each resembling a cluster of tough strings. The slightest breeze moves them, and they bob around against each other and the small branches in an odd sort of way. After so much threshing that they can hold no longer, the little nuts become loosened and begin to drop off a few at a time. Certain birds eat a few and loosen others, which escape. The illustration shows some of these nuts, each supplied with a ring of bristles about the base, which acts as a parachute to permit the wind the easier to carry them for some distance

before falling, or to drift them on the surface of the snow or ice.

23. Seeds that tempt the wind by spreading their sails.--On low lands in the cool, temperate climate of Europe, Asia, and North America, is a common plant here known as great willow-herb, a kind of fireweed (Epilobium angustifolium). There are several kinds of fireweeds. This one grows from three to five feet high, and bears pretty pink flowers. In mellow soil the slender rootstocks spread extensively, and each year new sprouts spring up all around, six to eight feet distant. Below each flower ripens a long, slender pod, which splits open from the top into four parts, that slowly curve away from a central column. The apex of each seed is provided with a cluster of white silky hairs nearly half an inch long.

The tips of the hairs stick slightly to the inside of the recurved valves, some hairs to one valve, and often others to the adjacent valve, thus spreading them apart with the seed suspended between. Four rows of the seeds are thus held out at one time. Often not over half, or even a tenth part, of the seeds are well developed, yet the silky hairs are present and float away in clusters, thus helping to buoy those that are heavy. This is a capital scheme, for when the pods are dry and unfurled, they silently indicate to the slightest breath of air that they are ready for a flight, and it doesn't take much to carry them for a long distance. As an active boy delights to venture again and again over thin ice on a shallow pond in the pasture, half fearing, yet half hoping, that he may become a hero by breaking through and escaping, so likewise many of these seeds and seed-like fruits spread themselves out, as if to tempt the wind to come along and attack them.

The twin fruits of the parsnip and some of its near relatives are light and thin and split apart, each holding on lightly to the top of a slender stem. In this position they are sure to be torn off sooner or later. Somewhat after the manner of the willow-herb behave the pods and seeds of willows, poplars, milkweeds, Indian hemp, and cotton.

24. Why are some seeds so small?--Do you know why so many kinds of

plants produce very small and light seeds? Would it not be better if they produced fewer and larger seeds, which would then be stronger and better able to grow under adverse conditions? But a large number of small seeds cost the plant no more effort than a small number of large ones, and the lighter and smaller the seeds and the more there are of them, the better their chances for distribution, especially for long distances. The minute size of spores of most of the fungi are given as reasons why so many of them are so widely distributed.

Why is a boy or man of light weight chosen to ride the horse on the race track? That the animal may have less weight to carry and thereby use his surplus strength in making better time. The less weight the parachute of the seed of the willow-herb has to carry, the greater the chances for success in making a long journey. Of the willow-herb it takes one hundred seeds to weigh a milligram, including the hairs attached to them, and it would take thirty thousand to weigh as much as an ordinary white bean.

25. Seeds with parachutes.--Many years ago large portions of Huron and Sanilac counties of eastern Michigan were swept by a fire so severe that the timber was all killed. Fifteen years later the woody growth consisted mostly of willows, poplars, and birches. The seeds of all kinds of willows and poplars are very light, and are produced in immense quantities. Like those of the great willow-herb, they are beautifully constructed for making long journeys through the air--a fact that explains the frequency of these trees in burned districts. A considerable number of seeds and fruits grow with a parachute attached at one end, not to prevent injury by falling from the tree top, but to enable the wind to sustain and transport them for a longer distance.

26. A study of the dandelion.--In spring the dandelion is almost everywhere to be found; every one knows it--the child to admire, the gardener to despise. From each cluster of leaves spreading flat in the grass come forth several hollow stems, short or tall, depending on the amount of sunshine and shade. Each stem bears, not one flower, but a hundred or more small ones. Around and beneath each yellow cluster are two rows of thin, green, smooth scales

(involucre).

The short outer row soon curls back, as though for rest or ornament, or for watching the progress of the colony above; but the inner row has a very important duty yet to perform in guarding the large family within. At night, or in daytime, if the day be wet, the long scales press like a blanket closely about the flowers, and do not permit them to come out; but when the sun is bright, it shrinks the outer side of these scales, which then curl apart, leaving the yellow flowers ready for bees to visit or boys to admire and study. For several days the flowers of a head blossom in succession, each night to be snugly wrapped by the scales, and the next day to be again left open, if the weather be fine. After each flower in turn has been allowed to see the light, and after all have been crawled over by bee and wasp to distribute the yellow pollen that seeds may be produced, there is nothing else to do but patiently wait for a week or two while receiving food from the mother plant to perfect each little fruit and seed. During all this period of maturing, day and night, rain or shine, the scales hold the cluster closely; the stem bends over to one side, and the rain and dew is kept from entering. After a while, on some bright morning, the dandelion stalk is seen standing erect again, and is probably surrounded by many others in a similar position. The dry air shrinks the outside of the scales, and they turn downward; the circle of feathers at the top of the slender support attached to the seed-like fruit below spreads out, and the community, which now looks like a white ball of down, is ready for a breeze. The feathery top is now ready to act as a parachute, and invites the wind to catch up the whole and float it away. If there is no breeze, the moist air of night closes the outer scales; each of the feathery tips closes, and all are secure till the next bright day.

Of a like nature are fruits of thistles, fireweed, prickly lettuce, sow thistles, scabiosa, valerian, cat-tail flag, cotton grass, some anemones, smoke tree, virgin's bower, and some of the grasses.

27. How the lily sows its seeds.--Ripened pods of lilies usually stand straight up on a stiff, elastic stem; beginning at the top, each one slowly splits into

three parts, which gradually separate from each other. Why do they not burst open all of a sudden, like pea pods, and shoot the seeds all about and have the job done with? Or why does not the pod burst open at the lower end first, instead of the upper?

Observe that the three opening cells are lashed together loosely with a latticework. No slight breeze can dislodge the seeds, but just see how they behave in a good gale! The elastic stems are swayed back and forth against each other, and some of the upper seeds are tossed out by the wind that passes through the lattice, and at such times are often carried to some distance. The seeds at the top having escaped, the dry pods split down farther and still farther and open still wider, till the bottom is reached. As the seeds are not all carried away the first or even the second time, and as succeeding breezes may come from different directions, it is thus possible for the lily to scatter its seeds in all directions.

The seeds of the lily are flat, very thin, and rather light, not designed to be shot out like bullets, but to be carried a little way by the wind; the pods are erect, and open at the top, that the seeds need not escape when there is no wind or unless some animal gives the stem a strong shake. The latticework was made for a purpose, and the gradual opening of the pods prevents the supply from all going in one direction or in one day, for a better day may arrive. The student will look for and compare the following: Iris, figwort, wild yam, catalpa, trumpet-creeper, centauria, mulleins, foxglove, beardtongue, and many other fruits.

28. Large pods with small seeds to escape from small holes.--The large ripe pod of the poppy stands erect on a stiff stem, with a number of small openings near the top. The seeds are nearly spherical, and escape, a few at a time, when the stem is shaken by the wind or some animal, thus holding a reserve for a change of conditions. Here is an illustration of ripe pods of a bellflower, Campanula turbinata, nodding instead of erect.

The small holes are still uppermost, but to be uppermost in this case it is necessary for them to be at the base of the pod.

29. Seeds kept dry by an umbrella growing over them.--When mature, the apple of Peru, Nicandra, keeps every dry bursting fruit covered with a hood, umbrella, or shed, so that seeds may be kept continually dry and may be spread with every shake by the wind, or by an animal, in rainy weather as well as in dry.

In the words of Dr. Gray, "The fruit is a globular dry berry, enclosed by a five-parted, bladdery inflated calyx." The margins of the lobes of the calyx curl upwards and outwards as the berry hangs with the apex downward.

The berry is as large as one's thumb, and when ripe, bursts open irregularly on the upper side as it hangs up under the calyx. As the covering of the pod opens more and more, a few seeds at a time may be rattled out by wind or animal. The numerous large and light fruits, with calyx surrounding them, are each supported on a nodding stem, stiff and elastic, which gives the wind a good chance to sway them about. Water does not seem to get into the berries even when they are torn open, for when it is poured over the branches it rolls off the calyx roof as freely as from a duck's back. The fruits of Physalis are apparently kept dry in a manner similar to the apple of Peru, although when first mature they are soft and juicy, considerably like a ripe tomato.

30. Shot off by wind or animal.--The calyx of sage, bergamot, and most other mints, remains dry and stiff, as a cup to hold one to four little round nutlets as they ripen. The figure shows two of these in section, as they are attached to the main stem of the plant, or one of its branches. Observe the direction taken by the upper and by the lower points of the calyx. When dry, the plant behaves somewhat as follows: when the wind jostles the branches against each other, or when an animal of some kind hits the plant, this movement causes many of these cups to get caught; but the elastic stem comes suddenly back to its place, and in so doing flips a nutlet or more from its mouth one to six feet, somewhat as a boy would flip a pea with a pea-shooter.

In our garden, July 2, when plants of sage, Salvia interrupta, were ripening their fruit, we found it difficult to collect any seeds, but seedlings were observed in abundance on every side of the plant, some to the distance of six feet. Plants dispersing seeds in this manner have been called catapult fruits. Examine ripening fruits of blue curls, pennyroyal, germander, balm, horehound, dittany, hyssop, basil, marjoram, thyme, savory, catmint, skullcap, self-heal, dragon's head, motherwort, and various dry fruits of several chickweeds.

31. Seed-like fruits moved about by twisting awns.--Most of the grains of grasses are invested with glumes, or chaff, and a considerable per cent of the chaff has awns, some of which are well developed and some poorly developed. The distribution of such grasses depends on several agents--wind, water, and animals. The chaff and awns of all are hygroscopic; that is, are changed by differences caused by variation of moisture in the air. Sweet vernal grass, tall oat grass, holy grass, redtop, animated or wild oats, blue-joint, and porcupine grass are among them. When mature, the grain and glumes drop off, or are pushed off, and go to the ground. When moist, these awns untwist and straighten out, but when dry they coil up again; with each change they seem to crawl about on the ground and work down to low places or get into all sorts of cracks and crevices, where the first rain is likely to cover them more or less with earth, after which they are ready for growth.

32. Grains that bore into sheep or dogs or the sand.--Porcupine grass, Stipa spartea, grows in dry soil in the northern states, but more particularly on the dry prairies of the central portion of the United States. This grass, when ripe, has a very bad reputation among ranchmen for the annoyance the bearded grain causes them. The grains are blown into the stubble among grasses with the bearded point down, sticking into the soil. The first rain or heavy dew straightens out the awns, which are twisted again as they dry. The bearded point works a little farther with each change, and after twisting and untwisting a number of times it gets down three or four inches into the sand, often to moisture, where the awns decay and the grain germinates. Here is an admirable scheme for moving about and for boring into the ground. But

this is not all. The grains are quick to catch fast to clothing, as people move among the plants, and they are admirably fitted for attaching themselves to dogs and sheep, which they annoy very much. These animals transport the grains for long distances. The twisting and untwisting of the awns enable the grain to bore through the fleeces, and even to penetrate the skins and make wounds which sometimes cause the death of the animal. Examine also seeds of pin clover, Alfilerilla, which is becoming abundant in many parts of the world.

33. Winged fruits and seeds fall with a whirl.--The large fruit of the silver maple falls in summer. As these trees are most abundant along the margins of streams, the fruit often drops into the water and is carried down stream to some sand drift or into the mud, where more sand is likely to cover them. Thus sown and planted and watered, they soon grow and new trees spring up. But in many instances a strong breeze, sometimes a whirlwind, has been seen to carry these mature fruits from the tree to a distance of thirty rods.

A thin sheet of paper descends more slowly than the same material put in the form of a ball. On the same principle, many seeds and fruits are flattened, apparently for a purpose; not that they may be easily shot through the air by some elastic force, not to increase their chances for attachment to animals, but to enable the wind to sustain them the longer and carry them farther. Some seeds and dry fruits are said to have wings, with the general understanding that they are by this means better fitted to be sustained in air. We shall find that all or nearly all flattened seeds and dry fruits, also winged seeds and fruits, are one-sided, unbalanced, and more or less twisted; consequently, in falling to the ground they whirl about, and are thus kept much longer in the air than they would be if shaped more like a winged arrow. Even the wings on the fruit of some of the ashes are twisted, though many of them are flat. Experiments with these things are sure to interest inquisitive children, or even older persons, when once started right; they are likely to prove as interesting as flying kites, skating, fishing, or coasting on the hillside. Try experiments with seeds of catalpa, trumpet-creeper, wild yam, pine, spruce, arbor vit? and fruits of maple, box elder, birch, hop tree, blue beech, ailanthus, ash, tulip tree,--in fact, anything of this nature you can find, whether the name is familiar or not. No two of them will behave in all respects alike.

34. Plants which preserve a portion of their seeds for an emergency.--Many a great general or business man has learned by experience and observation that it is usually unwise to exhaust all resources in one effort. If possible, he always plans to have something in reserve for an emergency--a loophole for escape from difficulty. We have seen in many instances that plants are endowed with the same trait. This is well illustrated by the way in which the jack-pine, Pinus [Banksiana] divaricata, holds in reserve a portion of its seeds, to be used in case the parent trees are killed by fire. In 1888 I made a study of this tree as it lives on the sandy plains of Michigan. The tree is often killed by fire, and never sprouts from the stump, as do oaks, willows, cherries, and most other trees. The jack-pine grows readily and rapidly from seed dropped on the sand, and begins to bear cones and seeds in abundance while it is yet only a few years old, perhaps as young as five years in some instances. The cones open slowly to liberate their seeds, some of them only after months or even years, and in some cases they never open at all. I have seen cones containing good seeds that had been nearly grown over by the tree. Dry weather, the dryer and hotter the better, causes many of these stubborn old cones to open their scales and allow the seeds to escape. What can be the advantage in cones of this nature? Let us see. A brisk fire passes over the ground at irregular intervals, usually of from one to ten years; it licks up all dry leaves and sticks, and kills the pine trees and all else above ground. The soil and the trunks of trees are blackened, and by lack of reflection the heat of the sun is rendered more intense; besides, the heat of the fire acts slowly on the unburned cones as they are left on the dead trees. By the time the quick hot fire has passed over, the cones have slowly opened and begun scattering seeds on the vacant and newly burned ground, at a time when there is the best possible chance for them to grow. I picked a few unopened cones which, according to my judgment, were from two to four years old. They were placed under glass in a dark sheet-iron dish and exposed to the sun. The extra heat caused the cones to open; many seeds were obtained and

sown, and in five days they began to come up, 95 per cent germinating. From the same tree I selected at the same time older cones, which I believe to be from four to six years old at least. From these, 225 seeds were sown, 191 of which germinated--about 85 per cent.

CHAPTER VI.

PLANTS THAT SHOOT OFF THEIR SPORES OR SEEDS.

By numerous devices a large number of the lower plants send off their ripe spores with considerable force. Some call them sling fruits. One in particular, Pilobolus cristallinus, found about damp stables, I have observed to shoot black masses of spores to a spot on a wall six feet above the ground, with enough force to have carried them not less than twelve feet. When ripe and dry, the spores of most ferns are shot from the parent plant by a motion forcible enough not only to burst the sporangium, the vessel that contains the spores, but also to turn it inside out.

35. Dry pods twist as they split open and throw the seeds.--In December, while absent from home, I collected for future study some pods of the Chinese wistaria, and left them on my desk in the library for the night. The house was heated by a hot-air furnace. In the morning the pods were in great confusion; most of them had split and curled up, and the seeds were scattered all about the room. As usual the little daughter, an only child, was accused of spoiling my specimens, but she showed her innocence. A little investigation and a few experiments with some pods not yet opened explained the whole matter satisfactorily. The stout pods grow and ripen in a highly strained condition, with a strong tendency to burst spirally, the two half-pods being ready to coil and spring in opposite directions; when the valves can no longer hold together, they snap with a sharp noise and sling the heavy seeds, giving them a good send-off into the world. As a pair of birds build a nest, hatch eggs, rear their young, and then send them forth to seek their fortunes, so for months the mother plant had labored, had produced and matured seeds, which at last it scattered broadcast. Goethe, Kerner von

Marilaun, each independently, and very likely others, had an experience with ripe pods brought to a warm room very similar to my own. In many cases the ripe and drying fruits are "touched off" by wind jostling the branches or by animals passing among them; in the latter case there is a chance that a portion of the discharges will be lodged somewhere on the animal and be carried along with it.

36. A seed case that tears itself from its moorings.--The perennial phlox in cultivation distributes its seeds in the following manner: when ripe, the calyx becomes dry and paper-like, and spreads out in the form of a saucer. The thick-walled dry pistil opens from the top into three pieces with a snap, spreading open so far against the calyx that it is torn from the brittle attachment; away go the seeds, mingled with the fragments of the pistil, no longer of any use.

Fruits that sling their seeds are to be found in every neighborhood, and are first-class objects for the curious person to see and handle. Very fortunate is the girl or boy who is never fully satisfied with what he reads and sees pictured, but has a strong desire to learn how plants are made and how they behave. A considerable number of seed pods have been illustrated with notes in recent schoolbooks. Here are some of them: peas and vetches, and some kinds of beans, violets, balsams, wood sorrel, geranium, castor bean, some of the mustards and cresses and their cousins, Alfilerilla, richweed, Pilea, witch-hazel, and others. Each of those will well repay study, especially the fruit and seeds of oxalis. The witch-hazel bears a hard, woody, nut-like fruit, as large as a hazelnut; when ripe, the apex gaps open more and more, the sides pressing harder against each smooth seed, till finally it is shot, sometimes for a distance of thirty feet. The girl who has shot an apple seed or lemon seed with pressure of thumb and finger across a small room, can understand the force needed to shoot a seed but little heavier than that of the apple two or three times that distance.

CHAPTER VII.

PLANTS THAT ARE CARRIED BY ANIMALS.

With the frosts of autumn ripe acorns, beechnuts, bitternuts, butternuts, chestnuts, hickory nuts, hazelnuts, and walnuts are severed from the parent bush or tree and fall to the ground among the leaves.

37. Squirrels leave nuts in queer places and plant some of them.--Even before the arrival of frosts many of these are dropped by the aid of squirrels, gray and red, which cut the stems with their teeth. The leaves, with the help of the shifting winds, gently cover the fruit, or some portions of it, and make the best kind of protection from dry air and severe cold; and they come just in the nick of time. Dame Nature is generous. She produces an abundance; enough to seed the earth and enough to feed the squirrels, birds, and some other animals. The squirrels eat many nuts, but I have seen them carry a portion for some distance in several directions, and plant one or two or three in a place, covering them well with soil. It may be the thought of the squirrel-- I cannot read his thoughts--to return at some future time of need, as he often does. But in some cases he forgets the locality, or does not return because he has stored up more than he needs; or in some cases the squirrels leave that locality or are killed; in any such case the planted nuts are not disturbed. At all events, some of the nuts--one now and then is all that is needed--are allowed to remain where planted. In this way the squirrel is a benefit to the trees and pays for the nuts he eats. He has not lived in vain, for he is a tree planter and believes in arboriculture. His arbor day comes in autumn, and he needs no message from the governor to stimulate him to work.

After some red squirrels had been given black walnuts, a member of my family saw them hide the nuts in all conceivable places, and in some instances place them above a cluster of small branches of a tree for support where three or more twigs spread from nearly the same place. Here the nuts, one in a place, were left till perhaps shaken to the ground by a severe wind or by some other cause. In one winter, without hunting for them, six to ten places were found in one neighborhood of Michigan, where something had placed a single walnut or acorn in the forks of small branches. In some cases a severe wind could have dislodged the nut.

On February 18, 1897, I found a single black walnut held by small branches of a red oak.

The oak was an inch and a half in diameter, and the nut was about six feet from the ground. The nearest bearing tree was fully three hundred long steps distant. We can imagine that, through fright or other causes, a squirrel might be suddenly interrupted while carrying nuts, and might then drop them to the ground, where later a tree would be started.

38. Birds scatter nuts.--The work of birds in scattering seeds and fruits has long been recognized.[3]

[Footnote 3: In the fall of 1897, Prof. C. F. Wheeler saw a blue jay fly from a white oak tree with an acorn in its mouth. The bird went to the ground four or five rods distant and crowded the acorn into the soil as far as it could, covering the spot with a few leaves. A member of my family saw a blue jay leave half of a black walnut in the forks of several small branches.]

Some friends of mine collected a quantity of hazelnuts, while yet the green husks enclosed the nuts, and placed them near the house to dry. At once they were discovered by a blue jay, which picked out a nut at a time, flew away, held the nut between its toes, cracked it from the small end, and ate the contents. In this operation a number of nuts slipped away and were lost. But it seems that all were not eaten, for the next season half a dozen or more hazel shoots came up, and to-day a new patch of hazel bushes is growing in the yard. Doubtless many acorns are carried from place to place and dropped in an aimless way by woodpeckers, blue jays, and crows; also beechnuts by these birds, and by nuthatches, and by pigeons, before the latter became nearly extinct. Woodpeckers and blue jays place beechnuts and small acorns in the crevices of bark on standing trees. If left there very long, the nuts will become too dry to grow, but in the act of transporting them some of the nuts may be accidentally dropped in various places.

39. Do birds digest all they eat?--To determine whether seeds would lose their vitality in passing through the digestive organs of birds, Kerner von Marilaun fed seeds of two hundred and fifty different species of plants to each of the following: blackbird, song thrush, robin, jackdaw, raven, nutcracker, goldfinch, titmouse, bullfinch, crossbill, pigeon, fowl, turkey, duck, and a few others; also to marmot, horse, ox, and pig, making five hundred and twenty separate experiments. As to the marmot, horse, ox, and pig, almost all the fruits and seeds were destroyed. From the ox grew a very few seeds of millet, and from the horse one or two lentils and a few oats; from the pig a species of dogwood, privet, mallow, radish, and common locust. Under ordinary conditions, no seed was found to germinate after passing through the turkey, hen, pigeon, crossbill, bullfinch, goldfinch, nutcracker, titmouse, and the duck. Ravens and jackdaws passed without injury seeds of stone fruits and others with very hard coats. Of seeds that passed through the blackbird 75 per cent germinated, 85 per cent in the case of the thrush, 80 per cent in the case of the robin.[4]

[Footnote 4: It should be noted that the blackbird here mentioned is not the same as either of our blackbirds, but a thrush much like our robin; that the robin mentioned is a ground warbler nearly related to our bluebird. It should also be noted that jackdaws, ravens, thrushes, and probably many others eject thousands of seeds by the mouth for one which passes through the intestines.]

40. Color, odor, and pleasant taste of fruits are advertisements.--In summer, buds are formed on bushes of black raspberry, blossoms appear, and these are followed by small, green, and bitter berries, which hardly anything cares to eat. They grow slowly, become soft and pulpy, and finally good to eat. How is bird or boy or girl to know where they are and when they are fit to eat? The plant has enterprise and has displayed two want advertisements by painting the berries first dark red, and then dark purple, when they are good to eat. But is the plant made expressly to produce berries, just to feed birds and children? If that be all, why are seeds formed in the berries in such large

numbers? No! They produce berries that contain seeds, and from these seeds are to grow more bushes. Then why should not the berries always remain bitter or hard, so that nothing would touch them? If we may say so, the plant produces sweet and showy berries on purpose to be eaten, that the seeds may be carried away. What becomes of the seeds? Each one is enclosed in a hard, tough covering, which protects it from destruction in the stomachs of many birds and some other animals. The seeds are well distributed by the animals that eat the berries. The brilliant colors of ripe berries say to bird and child: "Here we are; eat us, for we are good." The sweet pulp pays the birds for distributing the seeds, else they would not be so distributed. The seeds are as well provided for locomotion as the ticks, the mites, and the spiders, and when ready to go, the berries flaunt their colors to attract attention. You see, then, that although the old parent bush cannot change its place, young bushes grow from the tips of the branches, and seedlings spring up at long distances from their old homes.

Sparrows, finches, and similar birds in the winter eat and destroy seeds of grasses and weeds, while the same birds in summer and autumn eat bushels of blueberries, huckleberries, elderberries, raspberries, strawberries, and similar fruits, and distribute their unharmed seeds over thousands of acres, which otherwise might never support a growth of these species.

The downy woodpecker, among other things, devours berries of three kinds of dogwood, Virginia creeper, service berry, strawberry, pokeberry, poison ivy, poison sumac, stag-horn sumac, and blue beech.

The hairy woodpecker devours many of the above fruits, as well as those of spicebush, sour gum, cherries, grapes, blackberries. The flicker devours most of the fruits listed for the two woodpeckers named above, also hackberry, black alder, green brier, bayberries. A number of other woodpeckers possess habits much the same as the three above named. The cedar bird devours many species of hard-seeded fruits.

The various shades of red appear to good advantage among green leaves. As illustrations of such, we have the wintergreen, partridge berry, bush cranberry, bearberry, service berry, currant, holly, strawberry, red-berried elder, winter berry, honeysuckle, and many more. Where the leaves are liable to become red in autumn the berries are often blue. Of such, notice wild grapes, blueberries, and berries of sassafras, though the flowering dogwood has red leaves as well as red berries.

There is a reason for prickles on rosebushes. When ripe, rosehips are usually red or yellow, and thus attract birds which are fond of the fleshy portion outside; but the seed-like nuts are too hard and dry to suit their taste, and are rejected and sown in the vicinity, where the ripened hips are picked in pieces and eaten. Mice and red squirrels are also fond of the seed-like nutlets of roses, but seldom secure them from the bushes. Why, do you ask? Because the prickles were most likely placed on the rosebushes to prevent this very thing, and not to annoy the lover of flowers, or to prevent her from cutting what she needs.

41. The meddlesome crow lends a hand.--"One of the most industrious and persistent seed-transporting agencies I know of is that ubiquitous, energetic, rollicking, meddlesome busybody, the crow. I have seen crows gather by hundreds and have a regular powwow, a mass convention, where they seemed to discuss measures and appoint officers. At length they get through, and as they start to fly away many, if not all, will drop something. I have found these to be acorns, walnuts, hickory nuts, buckeyes, sycamore balls, sticks, eggshells, pebbles, etc. As a crow leaves an oak he will pluck an acorn, which he may carry five miles and light on a beech tree where something else will attract his attention, when he will drop the acorn and maybe pluck a pod of beechnuts and fly away somewhere else."--Prof. W. B. Barrows.

The number of seeds distributed by crows is enormous, and consists of many species, including poison ivy and poison sumac, wild cherry, dogwood, red cedar, sour gum, and Virginia creeper. The hard, undigested seeds are mostly expelled from the mouth in pellets, shown in the illustration, and germinate more promptly than those untouched by birds.

Bears are very fond of berries, and will scatter the seeds of service berries, elder berries, chokecherries, raspberries, and blackberries.

42. Ants distribute some kinds of seeds.--Ants are numerous, strong, skillful, and in suitable weather are always very busy. Their habits have been investigated, and it has been found that in some respects they are genuine farmers on a small scale. They have their slaves (not hired help); they feed their plant lice, remove them from place to place, and otherwise care for them, because the lice constitute one of the chief sources of their supply of sweet. They build roads and houses, and enjoy society after their fashion. They have use for certain kinds of seeds, portions or all of which they eat at once or carry to their homes. A number of persons in different countries and at different times have seen ants carrying seeds. Some young student of botany may have noticed along one side of the glossy seeds of the bloodroot a delicate, fleshy ridge, and wondered what could be its use. The answer can now be given with a good degree of confidence. The ants either eat this fleshy ridge at once, or, as more frequently happens, carry such seeds to their homes. The smooth seeds they do not eat, but cast them out of their nests after using the part they like; after being rejected the seed may stand a chance to germinate. The seeds cannot be carried so well unless this ridge, caruncle, be present. Other seeds of this nature are those of wild ginger, celandine, cyclamen, violet, periwinkle, some euphorbias, bellwort, trillium, prickly poppy, dutchman's breeches, squirrel-corn, several species of Corydalis, Seneca snakeroot, and other species of milkworts.

In his work on Vegetable Mold and Earthworms, p. 113, Darwin states that earthworms are in the habit of lining their holes, using seeds among other things, and that these sometimes grow. In this way the worms aid in spreading plants.

43. Cattle carry away living plants and seeds.--In Arizona, where cacti abound, Professor Toumey finds that many of them are broken in pieces by cattle, which eat a portion, while other portions often adhere to the legs or

noses and are carried from place to place. These fragments are usually capable of growing.

The unicorn plant, Martynia proboscidia, common in the southwestern portion of the United States, is sometimes seen in cultivation. When ripe, the fruit is hard, carrying two stout beaks with recurved tips. Experiments show it to be admirably adapted to catch on to the feet of sheep, goats, and cattle, or hold to the fleeces of the two former.

44. Water-fowl and muskrats carry seeds in mud.--Seeds and fruits of aquatic and bog plants that are floating, or in the mud of shallow water, are often carried by ducks, herons, swallows, muskrats, and other frequenters of such places, on their feet, beaks, or feathers, as they hastily leave one place for another. In this way seeds of water plantain, sedges, grasses, rushes, docks, arrowhead, pondweeds, duckweed, cat-tail flag, bur reed, bladderwort, water crowfoot, and many others are transported from one pond, lake, or stream, to another. In some cases enough of a living plant may be detached and carried away to keep on growing. Darwin found on the feet of some birds six and three-quarter ounces of mud, in which were five hundred and thirty-seven seeds that germinated. Mud may be carried on the feet of land animals as well as on aquatic animals, not only from ponds and bogs, but from the fields where seeds may have accumulated in the earth or washed down the slopes.

45. Why some seeds are sticky.--Some seeds and fruits are sticky; in some instances the mucilaginous substance is normally moist enough to adhere to anything that touches it, while in other cases it requires to be wetted before it will adhere. The seeds of flax, plantain, peppergrass, basil, sage, dracocephalum, groundsel, drop-seed grass, and many others less familiar, possess this peculiarity. The berries of some plants, when fully ripe, burst very easily when touched, and some of the seeds are then likely to adhere to animals and be carried away. Some berries of several plants belonging to the nightshade family have this peculiarity, as well as some of the cucurbits. When the outer covering of seeds of water lilies, arums, and others are

broken, the gummy secretion is very likely to adhere to the feathers, or fur, or feet of animals. A number of fruits, and even the upper fruit-bearing branches, have sticky glands with which to catch on to any passing object. Among these are some kinds of sedges, chickweeds, and catchflies.

The sticky substance on seeds and fruits not unfrequently serves another good turn besides enabling them to adhere to animals. The slime holds them to the spot where they are to grow, or it enables some to float or to sink in water, according to the amount of the mucilage.

46. Three devices of Virginia knotweed.--A perennial plant, four to five feet high, grows on low land, usually in the shade. It is Polygonum Virginicum, and so far without a common name, unless Virginia knotweed be satisfactory. It is a near relative of knot grass and smartweed and Prince's feather. The small flowers are borne on a long, elastic, and rather stiff stem, and each flower stalk has a joint just at the base. As this fruit matures, the joint becomes very easy to separate. It dries with a tension, so that, if touched, the fruit goes with a snap and a bound for several feet. The shaking produced by the wind jostling several against each other is sufficient to send off a number of ripe fruits in every direction. Like many other plants we have seen, this has more than one way of scattering seeds, and often more than two ways. Observe the slender, stiff beak, terminating in two recurved points. Let a person or some animal pass into a patch of these plants, and at once numerous fruits catch on wherever there is a chance, and some are shot upon or into the fleeces of animals, there to find free transportation for uncertain distances. Should there be a freshet, some of these fruits will float; or, in case of shallow currents after a rain, some of them are washed away from the parent plant. Any inquisitive person cannot fail to be pleased if he experiment with the plant when the fruit is ripe.

47. Hooks rendered harmless till time of need.--There are a number of rather weedy-looking herbs, common to woods or low land, known as Avens, Geum. They are closely allied to cinquefoil, and all belong to the rose family. The slender stiles above the seed-like ovaries of some species of Avens are

described as not jointed, but straight and feathery, well adapted, as we might suppose, to be scattered by the aid of wind; while others are spoken of as having, when young, stiles jointed and bent near the middle. In ripening, the lower part of the stile becomes much longer and stouter. When a whole bunch of pistils has drawn all the nourishment possible, or all that is needed, from the plant mother, the upper part of each stile drops off, leaving a sharp, stiff hook at the end. At this time each pistil loosens from the torus and can be easily removed, especially if some animal touch the hooks. To help in holding fast to animals, there are a number of slender hairs farther down the stile, which are liable to become more or less entangled in the animal's hair, fur, wool, or feathers. Even in the small number of plants here noticed, we have seen that scarcely any two of them agree in the details of their devices for securing transportation of seeds. I know of nothing else like the Geum we are now considering. When young and green, the tip of each hook is securely protected by a knob or bunch, with a little arm extending above, which effectually prevents the hook from catching on to anything; but, when the fruit is ripe, the projecting knob with its little attachment disappears. The figures make further description unnecessary. To keep the plow from cutting into the ground while going to or from the field, the farmer often places a wooden block, or "shoe," over the point and below the plow. Sometimes we have known persons to place knobs of brass or wood on the tips of the sharp horns of some of their most active or vicious cattle, to prevent them from hooking their associates or the persons having them in charge. Nature furnishes the points of the young fruits of some species of Avens with knobs, or shoes, for another purpose, to benefit the plants without reference to the likes or dislikes of animals.

[Illustration: FIG. 56.--The pistil of Avens in three stages of its growth.]

48. Diversity of devices in the rose family for seed sowing.--All botanists now recognize plants as belonging to separate families, the plants of each family having many points of structure in common. Among these families of higher plants, over two hundred in number, is one known as the rose family. Notwithstanding their close relationship, the modes of seed dispersion are

varied. The seeds of plums and cherries and hawthorns are surrounded by a hard pit, or stone, which protects the seeds, while animals eat the fleshy portion of the fruit. When ripe, raspberries leave the dry receptacle and look like miniature thimbles, while the blackberry is fleshy throughout. The dry, seed-like fruits of the strawberry are carried by birds that relish the red, fleshy, juicy apex of the flower stalk.

Each little fruit of some kinds of Avens has a hook at the apex, while in Agrimony many hooks grow on the outside of the calyx and aid in carrying the two or three seeds within. Plants of some other families illustrate the great diversity of modes of dispersion as well as the roses.

49. Grouse, fox, and dog carry burs.--To the feathers of a ruffed grouse killed in the molting stage, early in September, were attached fifty or more nutlets of Echinospermum Virginicum Lehm.

A student tells of a tame fox kept near his home, on the tail of which were large numbers of sand burs, and a smaller number on his legs and feet. Another student has seen dogs so annoyed by these burs on their feet that they gave up all attempts to walk.

Many wild animals unwillingly carry about such fruits, and after a while most of them remove what they can with claws, hoof, or teeth. Many of these plants have no familiar common names, but who has not heard of some of these? enchanter's nightshade, bedstraw, wild liquorice, hound's tongue, beggar-ticks, beggar's lice, stick-tights, pitchforks, tick-trefoil, bush clover, motherwort, sand bur, burdock, cocklebur, sanicle, Avens, Agrimony, carrot, horse nettle, buffalo bur, Russian thistle. Besides these, a very large number of small seeds and fruits are rubbed off and carried away by animals. Some of these stick by means of the pappus, as, for instance, the dandelion, thistle, prickly lettuce; others by means of hairs on the seed, such as those of the willow-herb and milkweeds and willows; or by hairs on the fruit, as virgin's bower, anemone, cotton grass, and cat-tail flag. These last named are apparently designed to be wafted by the wind, but they are ever ready to improve any other opportunity offered, whether it be by water or by clinging to passing animals.

50. Seeds enough and to spare.--In producing seeds nature is generous, often lavish. Most seeds are eaten by animals, or fall in places where they cannot germinate and produce plants, or fall in such numbers that most of them in growing are crowded and starved to death. A very small proportion fall on good ground, and succeed in becoming fruiting plants. A large plant of purslane produces one million two hundred and fifty thousand seeds; a patch of daisy fleabane, three thousand seeds to each square inch of space covered by a plant. The genuine student will not be satisfied till he has selected several different kinds of plants and counted, or estimated, the number of seeds produced by each, or the number of seeds furnished to the area covered by one or by several plants.

CHAPTER VIII.

MAN DISPERSES SEEDS AND PLANTS.

In describing the various means by which plants are dispersed, people are very likely not to mention the aid supplied by man, or to speak of his efforts as artificial or unnatural, forgetting for the time that man so far appears to be the crown of earthly existence, and that his works are a necessary part of a complete world.

51. Burs stick to clothing.--Late in summer or in autumn, who is there who has not returned from a walk along the river or from a tramp through thickets or the open woods, to find large numbers of half a dozen kinds of seed-like fruits sticking to his clothes? When ripe, these fruits usually separate from the parent plant very easily, by a joint or brittle place well provided for in the early part of the season. In pursuing your way you rub off a portion of these fruits, and at the end of the journey, or before, you sit down in some comfortable spot and deliberately pick off the unwelcome stick-tights. At such times you have been the means of transporting seeds, and you have left

them scattered about ready to grow. If you ever were so fortunate as to live on a farm, you must have seen your father or his hired help carefully look about the field or the wood lot and remove all the bur-bearing plants that could be found before turning in his flock of sheep or the colts and cattle; for if this were not done, he knows that hair and mane will surely be disfigured, and that the wool will be rendered unsalable. In removing the weeds he defeated the plans of Nature in her devices for sowing seeds.

The agency of man in the distribution of plants exceeds in importance that of all other means combined. He buys and sells seeds and plants, and sends them to all parts of the habitable globe. He exterminates many plants in large areas, and substitutes in large measure those of his choice. Mixed with seeds of grasses, clovers, or grains, he introduces many weeds and sows them to grow with his crops.

L. H. Dewey, in the Yearbook of the Department of Agriculture for the year 1896, p. 276, says: "Cockle seeds are normally somewhat smaller than wheat grains. In some parts of the northwest, where wheat for sowing has been cleaned year after year by steam threshers, all the cockle seeds except the largest ones have been removed, and these have been sown until a large-seeded strain has been bred which is very difficult to separate from the wheat." For illustration, some years ago I purchased of a dealer in Michigan a small quantity of what was being sold on the market as seed of red clover; this specimen contained 40 per cent of seeds of rib-grass or narrow-leaved plantain.

Man introduces some seeds of weeds with unground feed stuff. He introduces some with barnyard manure drawn from town. He gets some in the packing of nursery stock, crockery, baled hay and straw. For example, in 1895, baled hay from Kansas or that vicinity examined at the Missouri Agricultural College was found to contain fifteen species of weeds. Others from the west were examined in Michigan and found to contain much foul stuff. Some are carried from farm to farm by wagons, sleighs, or threshing machines; or they are spread by plows, cultivators, and harrows. A few are introduced to grow for ornament or food, and afterwards spread as weeds. A number have been shipped to distant lands in the earth of ballast, which is often unloaded and reloaded at wharves where freight is changed. They are carried along the highway, strung along the towpath of canals, or are carried in the trucks or in the cars of railroads. They are imported and exported around the world in fleeces of wool. They float down irrigating ditches from farm to farm, and with the water are well distributed.

52. Man takes plants westward, though a few migrate eastward.--So far as man's agency is concerned, the direction for plant migration is generally westward, in the course taken by himself. In case of two hundred kinds of weeds named by the United States Department of Agriculture, one hundred and eight species are of foreign origin. Three notable samples of weeds in the United States have gone from the west to the east, carried in seeds of grasses or clovers. These are Rudbeckia hirta, Artemisia biennis, Plantago aristata. To these Mr. Dewey adds buffalo bur, Solanum rostratum, squirreltail, Hordeum jubatum, false ragweed or marsh elder, Iva xanthifolia, Franseria hookeriana, alfalfa dodder, Cuscuta epithymum.

Above I have barely mentioned a few of the methods by which man is an unwilling agent in distributing plants. Large volumes could be filled with statements of man's more or less carefully planned attempts to transport seeds and living plants from one part of the world to another.

CHAPTER IX.

SOME REASONS FOR PLANT MIGRATION.

53. Plants are not charitable beings.--Man uses to his advantage a large number of plants, but there appears to be no evidence that the schemes for their dispersion were designed for anything except to benefit the plants themselves. The elegant foliage and beautiful flowers, the great diversity of attractive seeds and fruits, all point to plants as strictly selfish beings, if I may so use the term; and not to plants as works of charity, to be devoured by

animals without any compensation. By fertilizing flowers, by distributing plants, and by other helpful acts, animals pay for at least a portion of the damage they do.

By an almost infinite number of devices, we have seen that seeds and fruits flee from the parental spot on the wings of the wind, float on currents of ocean, lake, and river. They are shot by bursting pods and capsules in every direction. With hooks, barbs, and glands they cling to the covering of animals. Allured by brilliant colors, birds and other animals seek and devour the fruits of many plants, the seeds of which are preserved from harm by a solid armor; these seeds are then sown broadcast over the land, ready to start new colonies. Nuts are often carried by squirrels for long distances, and there securely buried, a few in a place. By a slow process, which, however, covers a considerable space, in a few years many plants send forth roots, rootstalks, stolons, and runners, and thus increase their possessions or find new homes.

54. Plants migrate to improve their condition.--The various devices by which plants are shifted from place to place are not merely to extend and multiply the species, and reach a fertile soil, but to enable them to flee from the great number of their own kind, and from their enemies among animals and parasitic plants. The adventurers among plants often meet with the best success, not because the seeds are larger, or stronger, or better, but because they find, for a time, more congenial surroundings. We must not overlook the fact, so well established, that one of the greatest points to be gained by plant migration is to enable different stocks of a species to be cross fertilized, and thereby improved in vigor and productiveness.

55. Fruit grown in a new country is often fair.--Every horticulturist knows that apples grown in a new country, that is suited to them, are healthy and fair; but, sooner or later, the scab, and codling moth, and bitter rot, and bark louse arrive, each to begin its particular mode of attack. Peach trees in new places, remote from others, are often easily grown and free from dangers; but soon will arrive the yellows, borers, leaf curl, rot, and other enemies. For a few years plums may be grown, in certain new localities, without danger

from curculio, or rot, or shot-hole fungus. It has long been known that the nicest way to grow a few cabbages, radishes, squashes, cucumbers, or potatoes is to plant a few here and there in good soil, at considerable distances from where any have heretofore been grown. For a time enemies are not likely to find them. I have often noticed that, while pear-blight decimated or swept large portions of a pear orchard, a few isolated trees, scattered about the neighborhood, usually remain healthy. The virgin soil of the Dakotas produced, at a trifling cost, healthy, clean wheat, but it was not long before the Russian thistle, false flax, and other pests followed, to contest their rights to the soil.

As animals starve out, in certain seasons when food is scarce, or more likely migrate to regions which can afford food, so plants desert worn-out land and seek fresh fields. As animals retreat to secluded and isolated spots to escape their enemies, so, likewise, many plants accomplish the same thing by sending out scouts in all directions to find the best places; these scouts, it is needless to say, are seeds, and when they have found a good place, they occupy it, without waiting for further instructions.

56. Much remains to be discovered.--"In this, as in other branches of science, we have made a beginning. We have learned just enough to perceive how little we know. Our great masters in natural history have immortalized themselves by their discoveries, but they have not exhausted the field; and if seeds and fruits cannot vie with flowers in the brilliance and color with which they decorate our gardens and our fields, still they surely rival them--it would be impossible to excel them--in the almost infinite variety of the problems they present to us, the ingenuity, the interest, and the charm of the beautiful contrivances which they offer for our study and our admiration."[5]

[Footnote 5: Flowers, Fruits, and Leaves, by Sir John Lubbock, p. 96.]

Frequent rotations seem to be the rule for many plants, when left to themselves in a state of nature. Confining to a permanent spot invites parasites and other enemies, and a depleted soil, while health and vigor are secured by frequent migrations. The more we study in detail the methods of plant dispersion, the more we shall come to agree with a statement made by Darwin concerning the devices for securing cross-fertilization of flowers, that they "transcend, in an incomparable degree, the contrivances and adaptations which the most fertile imagination of the most imaginative man could suggest with unlimited time at his disposal."[6]

Let no reader think that the topics here taken up are treated exhaustively, for if he will go over any part of this work and verify any observation or experiment, he will be sure to find something new, and very likely something different from what is here stated.

###

www.ingramcontent.com/pod-product-compliance
Lightning Source LLC
Chambersburg PA
CBHW070923180526
45168CB00005B/2121